PROFESSIONAL ENGINEER'S LICENSE GUIDE

PROFESSIONAL ENGINEER'S LICENSE GUIDE

**What You Need to Know and Do
To Obtain PE (and EIT) Registration**

Third Edition

Joseph D. Eckard, Jr., PhD, PE
Introduction by James T. Cobb, Jr., PhD, PE

Herman's Career Management Guides
Herman Publishing, Inc.
45 Newbury Street   Boston, MA 02116

Printed in USA
10 9 8 7 6 5 4 3 2 1

Library of Congress Cataloging in Publication Data

Eckard, Joseph D   , 1942-
   Professional engineer's license guide.

   (Herman's career management guides)
   Bibliography: p.
   Includes index.
   1. Engineers--Licenses-- United States. I. Title.
TA157.E26 1978   620'.0023   78-17249
ISBN 0-89047-020-0
ISBN 0-89047-019-7 pbk.

Contents

Author's Preface   vii

Acknowledgments   ix

Introduction: Trends in the Regulation
of the Practice of Engineering   xi

1 The Decision to Register   1

2 Registration Requirements   7

3 Registration Schedule   13

4 The Application   17

5 Structure of the Examinations   23

6 Preparing for the Examinations   33

7 Taking the Examinations   41

8 Completing Registration   47

9 Registration in More than One Jurisdiction   49

Appendix A   Registration Checklist   59

Appendix B   Exam Subject Area Preparation Checklist   61

Appendix C   Checklists of Equipment and Books to Take
to the Examinations   63

Appendix D   References   69

Appendix E   Addresses   79

Appendix F   Stated Positions of Some Engineering Societies
Regarding Registration   93

Appendix G   Sample Project Summary Pages   99

Index   105

Author's Preface

The purpose of this guide is to assist you in becoming a registered/licensed professional engineer (PE) or an engineer-in-training (EIT). It thus

- details the steps you must take
- describes what to expect at each step
- answers questions you probably will have
- provides references to necessary information
- presents successfully used techniques for organizing and executing your registration program.

Initial PE registration is not a simple matter; it requires a substantial investment of time. You must organize your program carefully so that you do not fail in your first attempt. This includes developing schedules for meeting the various deadlines and for preparing for the examinations, reviewing many subjects, working sample problems, working past examinations, and reviewing a variety of useful references. If you are thoroughly prepared, you will be confident in carrying out the somewhat protracted procedure of registration.

This guide is intended to be a unifying outline for you to use in conjunction with the references and other information it cites. It follows chronologically the actions you must take. An annotated list of references is given in Appendix D. These publications are extremely useful, and most are widely available in libraries and bookstores. You will probably use these and similar ones heavily.

If you follow the recommendations in this guide, you will encounter few, if any, surprises on your path to registration.

The Introduction by Dr. James T. Cobb, Jr., "Trends in the Regulation of the Practice of Engineering," is recommended as an excellent overview of the current status of licensing which will also alert you to possible significant changes ahead.

Acknowledgments

The author and publishers are grateful indeed to the many persons and organizations whose generous help was indispensable to this publication, but who are, of course, too numerous to allow individual mention. However, we must single out for special thanks a few whose contributions were especially valuable, who gave freely of their time and expert knowledge, and whose sole motivation was an unselfish commitment to the betterment of the engineering profession:

> *James T. Cobb, Jr.*, who provided the informative and important Introduction as well as invaluable advice;
> *Donald D. French*, former director of continuing professional education at Northeastern University and now president of the Institute for Advanced Professional Studies, whose assistance was vital;

We are grateful also to the American Institute of Chemical Engineers (AIChE) and the American Society of Civil Engineers (ASCE) for permission to quote in full in Appendix F their stated positions regarding registration.

Introduction Trends in the Regulation of the Practice of Engineering

By James T. Cobb, Jr., PhD, PE

Associate Professor of Chemical Engineering, University of Pittsburg.
Chairman, Commission on Certification, Engineers' Joint Council.
Chairman, Professional Legislation Committee, American Institute
of Chemical Engineers.
Representative of American Institute of Chemical Engineers to the
National Council of Engineering Examiners.

By the successful completion of the procedures described in this book you will be joining the growing number of registered "Professional Engineers." This number should continue to grow because of increasing public concern over the safety of manufacturing plants, the reliability of products of technology, and the quality of the environment.

I should like to explore here the climate surrounding the registration process today and what I see for the future.

First, it must be remembered that there are fifty-five boards of engineering registration in the United States alone, not to mention Canada and other countries. Each is responsive to a different legislature, has a different industrial setting, and is made up of different people. Although all fifty-five boards have banded together voluntarily to form the National Council of Engineering Examiners

(NCEE) and they continually strive for uniformity in their governing legislation and in their registration procedures, significant differences do exist among the boards and these probably will continue.

The factors which can affect registration in the future may be categorized by their tendency to either increase or decrease the necessity for registration and the amount of effort required to become and remain a registered Professional Engineer.

Trends Toward More Registration Activity

In addition to the public practice of engineering, which forms the basis for registration laws, other functions for which registered engineers are required are those specified by the regulatory agencies of the federal and state governments.

At the federal level, agencies such as the Nuclear Regulatory Commission (NRC), the Environmental Protection Agency (EPA), and the Occupational Safety and Health Administration (OSHA) require registration of certification in some form before one can perform such diverse tasks as operating a nuclear power plant, sealing a stack sampling report, or designing a ladder for use in an industrial plant.

Several states require registration to practice engineering before any political subdivision (e.g., a city or county) of the state (New York, for instance), or to take the witness stand in a civil suit involving technical matters (Alabama, for instance). These examples cover functions that one normally would expect to be covered by the various "exemption" clauses in many state registration laws. But, as you see, they not always are.

It appears likely at this writing that some sort of "Consumer Protection Agency" will be established by the U.S. Congress in the foreseeable future. Should this occur, there is a strong likelihood that regulations similar to those of EPA and OSHA will be published requiring registration for engineers in charge of the design and manufacture of many consumer products.

Such regulations of a consumer-oriented agency would considerably erode the so-called "manufacturer's exemption" which exists now in many state laws.[1]

---

1. A "manufacturer's exemption" provision in a registration law allows an engineer who is an employee of a manufacturer to practice without being registered. The rationale is that the manufacturer is a satisfactory judge of the employee's qualifications and should be given the responsibility to determine for the state the employee's competence.

In fact, perhaps in response to the same societal pressures which are leading to the call for a "Consumer Protection Agency," the NCEE has removed mention of a "manufacturer's exemption" clause from its "Model Law" which is used by many legislatures as a guide for revisions of state laws. As a result, the manufacturer's exemption has recently been removed from the Montana law. Other state legislatures undoubtedly will consider similar changes.

Trends Toward Less Registration Activity

While many state legislatures and federal agencies seem to be encouraging more regulation of engineers, others appear to be moving in the opposite direction.

At the federal level, officials within the Department of Justice and the Federal Trade Commission are concerned about possible monopolistic aspects of registration of all professionals which might result in a type of restraint of trade. Others, in the Equal Employment Opportunity Commission, are concerned about possible discriminatory features of the registration process whereby tests and education requirements might be structured to exclude individuals of certain minority groups from full professional practice.

At the state level, several legislatures have passed so-called "Sunset Laws," which require all state agencies, including the various registration boards, periodically to justify their continuation. Those failing to do so will automatically cease to exist after a specified date. This procedure was used in 1977, for the first time anywhere, in Florida. The results in that state may well set the trend for the others with similar "Sunset Laws."

Several other states recently have added members from the general public to their registration boards. Still others have pooled all registration boards into a single omnibus agency. The full effects of these changes are yet to be determined.

Technical Societies State Their Positions

The stated positions of some of the larger technical societies on the various issues of concern relating to registration are presented in Appendix F. You should find it valuable to know how the society for your particular specialty stands on these matters.

Trend Toward "Certification" by the Specialized Engineering Societies

A new recognition procedure has been introduced to several branches of engineering and related fields in recent years. This is called "certification," and it is similar to but not a substitute for the registration process. Certification is available either as an additional credential following registration (e.g., the program of the American Academy of Environmental Engineers) or as an alternative credential to registration (e.g., the program of the Society of Manufacturing Engineers).

A total of thirteen such "certification" programs have been developed and the number is expected to increase. Besides the two mentioned above, the other eleven include the American Institute of Plant Engineers (AIPE); the American Association of Cost Engineers (AACE); American Society of Safety Engineers (ASSE); the National Association of Corrosion Engineers (NACE); the American Society for Quality Control (ASQC); the Society of Packaging and Handling Engineers (SPHE); the American Production Inventory Control Society (APICS); the Standards Engineer's Society (SES); the Engineering Technology Certification Institute (ETCI); the Institute of Certifying Engineering Technicians (ICET); and the National Certification Commission in Chemistry and Chemical Engineering (NCCCE). Quasi engineering and technical certification programs are also provided by the

American Welding Society (AWS) and the American Society for Nondestructive Testing (ASNT).

One reason why these thirteen societies have begun certification programs is to provide peer recognition of a minimum level of knowledge and experience in the practice of the specialty, thereby better defining the specialty and qualified specialists.

A number of industrial firms have added specialty certification as a requirement for some positions and several government agencies are requiring that only certified specialists be permitted to carry out certain aspects of work contracted to industry. In addition, several of these disciplines have been added to the Principles and Practice portion of the state registration tests. However, it needs to be remembered and emphasized that certification is not a substitute for registration, which is conferred only by the individual states.

Trend Toward Recertification and Reregistration

Many of the societies' certification programs now in operation require recertification every three to five years. Requirements for recertification generally include various professional development activities such as continuing education, attendance at technical meetings, and home-study. If the minimum of such activity has not been maintained, the certification test must be retaken (or, in some cases, a portion of it).

Three state societies of professional engineers (Florida, New Jersey, and Wisconsin) have begun to offer programs for their members which are very similar to that type of recertification process whereby professional development activity is recognized. Such programs may spread further. For example, several major societies are investigating this form of member service, even though they are not at all interested in certification.

Many states have legislated reregistration procedures requiring professional development activities for a number of professions, principally in health-related fields. Until recently, the engineering profession was excluded from such procedures. In 1976 the Minnesota board of engineering registration was given the authority to establish reregistration requirements of this type. The enabling legislation did not require the Minnesota board to do so, and the board has stated clearly that it does not intend to establish a reregistration program.

In 1977, however, Iowa adopted a law which *re-*

quires the Iowa board of engineering registration to establish a reregistration procedure requiring professional development activities. The Iowa board has not yet completed the development of its program, but it will probably be similar to those used in recertification. Now that one state has passed a mandatory reregistration law of this type, other states are likely to pass similar laws. You should keep a very close watch on your state legislature in this area in particular.

The Decision to Register

What is Registration?

Registration or licensing (to be used synonymously) as a Professional Engineer is legal recognition by a state, the District of Columbia, or other jurisdiction of your qualifications to practice engineering. It is normally in one or more broad categories of engineering — mechanical, electrical, civil, nuclear, and so forth.

An applicant indicates on the application that discipline or branch of engineering in which he considers himself most proficient. The certificate of registration issued to a registrant may or may not indicate any specialty. This varies from state to state. Most states rely on your ethics to limit your practice to the areas of your experience and competence.

Who Needs to Register

Not all engineers need to be registered, although it is essential for some. Generally, those requiring it are carrying personal responsibility and authority for projects involving public safety. Those working on such projects from behind the corporate or governmental veil need not be registered, but a corporation engaged in certain engineering activities ultimately requires certification of its work by a PE, possibly a principal or an employee.

If you are now practicing as a professional engineer and are not licensed by your state, it is by virtue of exemptions from registration which the law permits for certain engineers.

Exemptions

Exempt persons typically include those working as subordinates or employees of licensed engineers, government officers and employees, employees of public utilities, and officers and employees of corporations working for the government or engaged in manufacturing or industrial processes. There is increasing pressure, however, for these exemptions to be dropped so that the stance taken by states toward engineering would be more in line with the position already taken toward other professions, such as medicine, law, and architecture, where there are no exemptions. Indeed, legislation is now being prepared in several states to require licensing of all persons engaged in professional engineering activities. (See Introduction.)

There are definite advantages to be gained by registration. (See Table 1.1.) The decision to register depends finally on your own circumstances and goals.

TABLE 1.1   **Advantages of Registration**

Increases professional flexibility and opportunity by opening more areas in which to practice

Permits advertising as an engineer

Permits participation as a principal in a professional corporation

Permits full participation in professional engineering societies with their attendant benefits (referrals, advertising, business assistance)

Gives your work legal authority

Often increases your credibility as an expert witness

Facilitates resorting to the courts to collect fees

Precludes your facing penalties or fines for practicing without a license

Gives some intangibles, such as added prestige, sense of responsibility, and sense of accomplishment

Often is a requirement for promotion within your company

Can You Register?

Once you have decided that you would benefit from registration, you must determine whether you are eligible. This is discussed in Chapter 2.

How Long Does Registration Take?

Next comes the question of when you will start your registration proceedings. If you meet the requirements and pass the exams, registration may take about ten months. If you do not meet the experience requirements (typically four years), you will have to defer full registration as a PE until they are met. You may — and in fact should — serve as an Engineer-in-Training (EIT) until you have gained the experience. Most states have EIT programs (Chapter 3 discusses this option and typical schedules).

Preparation

If you decide to proceed, you must concentrate first on completing the application and then on preparing for and taking the examinations. Chapters 4 through 7 cover these aspects. Chapter 8 covers the final steps you must take after you pass the exams.

Multiple Registration

You may find you need to register in more than one state. Multiple registration and national certification are covered in Chapter 9.

Motivation, Hard Work Needed

Your chances of fulfilling all the state requirements the first time around will be markedly improved by strong motivation to succeed and by developing your competence. You alone must supply the motivation and the hard work required to hone your competence. This guide will show you where you can most efficiently concentrate your effort and will build your confidence by showing you what to expect.

How to Start

Read through this guide completely before starting. Then proceed with the assurance that if you meet the education and experience requirements and if you prepare thoroughly for the examinations, you will be a registered PE withing a year (or you will have passed the initial phases by becoming an EIT).

Registration Requirements

2

**No National
Licensing**

Registration and licensing are powers reserved to the states; there is no national registration of engineers. Until recently the individual states had widely differing requirements for their licenses. The requirements are now becoming more uniform, however, and once you are registered in one state, registration in others is relatively simple. This is largely because most jurisdictions have adopted the uniform exams prepared by the National Council of Engineering Examiners (NCEE). With registration based on these exams, registration by comity (courtesy) is more easily established between the states. (Multiple registration is discussed in Chapter 9.)

**Typical
Requirements**

Your state may differ somewhat, but you will find that most have the following general requirements for registration as a PE:

- completing a rather extensive application, including character references from several PEs
- having one of several combinations of education and experience (typically an engineering degree and four years' experience)

- passing written (and possibly oral) examinations
- being over twenty-five years of age
- paying registration and examination fees.

Obtain Your Board's Package

One of your first actions must be to obtain the registration material from your state board for professional licensing and regulation. (See Appendix E for addresses.) Included will be the application form, instructions, and a copy of the state registration law. The requirements you must meet will be spelled out in the registration law.

Educational Requirements

Educational requirements and professional experience can be substituted for each other within certain limits. Typical combinations include (1) graduation from an engineering curriculum of four years or more approved by the Engineers' Council for Professional Development (ECPD) plus four years of professional experience; (2) graduation from a non-ECPD approved engineering curriculum of four years or more plus eight years of professional experience; (3) no engineering degree but twelve years of professional experience. The ECPD list of institutions with accredited curricula can be obtained directly from the ECPD or from the National Society of Professional Engineers. (See addresses in Appendix E.) The above combinations of education and experience are fairly closely followed;

however, the state board makes the final determination of whether an individual meets the minimum requirements.

Professional And Pre- professional Experience

You will have to exercise some thought in assessing whether your professional experience satisfies the requirements. Most states make the distinction between "professional" and "preprofessional" experience and you will be expected to perceive this distinction. Briefly, professional work is that which makes wide use of engineering principles, requires initiative, and demonstrates competence and good judgment. Preprofessional work is that which is routine or involves execution of work already designed by a professional engineer. The first six to twelve months' experience after receiving the BS degree is usually looked upon as preprofessional. Your application instructions will probably clarify this matter.

The EIT Option

If you do not meet the requirements for professional experience, you can still take the first half of the exams and, upon passing, serve as an engineer-in-training (EIT) until you gain the required experience.

EIT Requirements

The requirements for taking the EIT exam alone are typically rather minimal and include the four-year educational requirement, an abbreviated application form with references, and a reduced fee. The engineer-in-training does not command the legal authority of the professional engineer, but normally enjoys near equality with the PE in professional societies and their activities. Several states have changed the title of the EIT to Intern Engineer or something similar and other states are considering such changes. Some employers, when hiring a new engineering graduate, prefer that he or she already have EIT registration. It makes good sense for engineering students to take the EIT exam at or near the end of their senior year while course material is still fresh in mind. It is quite a chore to have to brush up on fundamentals after you have been out of school for several years.

The Examinations

EIT

PE

The examinations are typically written (as opposed to oral) and require a total of sixteen hours to complete. The first eight hours are devoted to the examination in Fundamentals of Engineering or the EIT exam. It covers a broad range of fundamental subjects which will be discussed in Chapter 5. The remaining eight hours of examination are for the examination in Principles and Practice of Engineering or the PE exam, which delves into a more specific area of specialization.

Timing

Some states permit a candidate to take the EIT and

PE exams on consecutive days (usually Saturday and Sunday).[1] Most, however, require that the EIT examination be passed before allowing the PE examinations to be taken, thus forcing at least a six-month interval between exams. Some states now give the two exams simultaneously, which also forces a six-month interval.

A great deal of your effort will be spent in preparing for these examinations. There is usually no way of avoiding them unless you have a vast amount of experience and can convince the board of your "eminence" in your field of engineering.

Age

If you do not meet the age requirement but feel otherwise qualified, you might consider registering in a nearby state not having a minimum age. Usually, though, anyone who has the required experience easily meets the age requirement.

Fees

The registration and renewal fees vary from state to state. Fees for registration range from $20 to $150, with $50 being typical. Those for renewal range from $6 to $30, with $15 typical. You might consider asking your employer to pay these fees, especially if they are high. He often will be willing to do so if your becoming a PE will enhance your value to him. Many companies have ongoing pro-

1. The NCEE will be recommending to the states that, starting in 1980, its exams be given on three consecutive days, Thursday, Friday and Saturday, the EIT on Thursday, the PE on Friday and both on Saturday.

grams by which they encourage employees to register. Contact your company's "professional development" officer or your personnel department.

Other Requirements

A few other requirements are specified by several states. Some states require U.S. citizenship or at least proof of intention to become a citizen. Some have residency requirements unless you are already registered in your home state. It is thus very important to obtain a copy of your state's registration law and to read carefully the sections on requirements.

Registration Schedule

3

Registration Takes Several Months

Obtaining full PE registration is not an overnight matter. It will probably require at least ten months. During about half of this time, you will be preparing application materials, preparing for the examinations, and taking the examinations. The last half will be spent waiting for your exams to be graded and for the results to come from your state board.

The schedules given here are based on the assumption that your state uses the NCEE uniform exams. Make any necessary adjustments if it does not.

Figure 3.1 shows an overall schedule, as well as the relative level of effort that the various phases will take. Note that completion of the application form is itself a fairly large task since so much information and so many people must be coordinated. Allow at least a month for this, or even better, two months. You can turn your application in early, but if you miss a submission deadline, you will face a six-month delay.

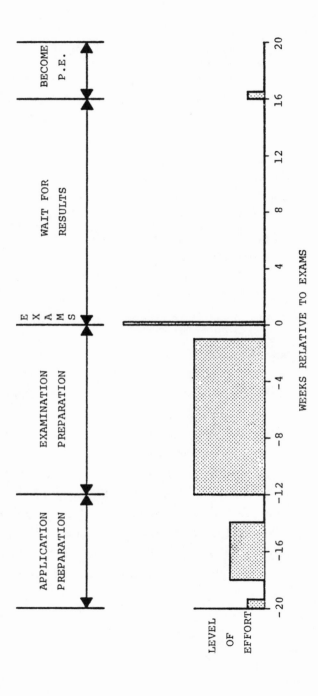

FIGURE 3.1   REGISTRATION PHASES AND LEVEL OF EFFORT

The exam preparation period is somewhat flexible. You can concentrate your study into seven nights each week for a short time or study more leisurely over a longer time. Once you get started, though, you will probably wish you had started earlier because of the large number of topics and subtopics you will have to review. Plan to spend about three months for your review. Two or three hours per night for two or three nights a week over this period should be adequate. If you have a family, you should discuss this schedule with them so that they realize the necessity of such a significant commitment of your time.

After the exams, you will have little to do (unless you fail them) except to complete the minor details discussed in Chapter 8. There are no deadlines for these but you will, of course, want to finish them quickly so that you can become a fully registered PE.

Take EIT as Soon as Possible

If you are a young engineer who has not yet accumulated the necessary professional experience to take the PE exams, you should take the EIT exams as soon as possible. The best time is just out of college or at the end of your senior year while you are still in command of the broad range of fundamentals in engineering, the sciences, and mathematics. Some schools, in fact, actively, explicitly prepare their students for the professional examinations.[1]

By taking the EIT exam separately, you will avoid the ordeal of two full consecutive days of examinations (where otherwise permitted) as well as the increased preparation effort.

In preparing for only the EIT examination, you may reduce the application preparation time considerably from that shown in Figure 3.1 since it may not encompass the detailed evidence of your professional experience. The examination preparation time will also be reduced, and the waiting time for the exam results will be only about two months.

Follow the Registration Checklist

Appendix A presents a registration checklist, which gives a more detailed schedule relative to the date of your exams. The exams are usually given in April and November. Fill in the actual dates when you find out from your state board exactly when they are. You should consult this schedule frequently and follow it.

---

1. At Georgia Institute of Technology, engineering students are required to pass the EIT examination in order to graduate.

The Application   4

The application for registration presents to the state board a general picture of you from which eligibility is determined. You should therefore prepare the application and supporting material with special care.

The complete application package that you submit can be divided into three parts: (1) the application form and fee; (2) verification of education; and (3) examples of work produced during the years listed under the experience requirements.

Application Form

The information requested on the application form itself is straightforward and usually includes

- personal data (name, address, age, social security number, etc.)
- recent photograph
- previous registration data
- professional society memberships
- educational experience record
- professional experience record
- names, addresses, and signatures of references

- your signature affirming that all information is true
- certification by a notary public.

Pick Your References Carefully

After you have filled in your personal, educational, and professional histories, some states require that you show the application to the persons who are serving as your references. They will read the application and sign it. Give some thought to selecting these references. Usually five are required, three or more of whom should be registered PEs (not necessarily in your state). These references will attest to your professional experience, your character, and your integrity. Therefore, choose references who know you well enough. Also, choose people who will respond promptly to the letter from your state board. You might even indicate to them the importance of a timely reply.

Notarize

Note the last item in the above list—notarization. Many states require this, and it is easy to overlook if you do not add it to your checklist.

Verification of Education

Verification of education should present no problems. Be sure not to omit any pertinent information. Obtain copies of graduation certificates and/or transcripts from all educational institutions you have attended, even for such peripheral training as business. Unless instructed to have these documents sent directly from the schools, you should obtain them yourself and include them with the rest of your application material. This ensures that you will be aware of any problems in getting them.

Your theses are additional evidence of your academic work. Be sure to include copies of the signed title page and abstract from each.

Other Training

You may want to include other training you have received to "flesh out" the profile you present to your state board. Such training, though not claimed as professional experience, may help swing situations of borderline experience in your favor. Consider reporting your training as a machinist, pilot, surveyor, welder, mechanic, seaman, electrician, or reactor operator; in computer programming, computer maintenance, law, business, drafting, or electronics; and for radio licenses.

Examples of Work

The examples of your professional work are a very important part of the application. If you are typical, you will have preserved copies of your papers, drawings, publications, reports, and so forth from each project on which you have worked. If you haven't, you should make an effort to obtain them because an application is immeasurably more credible when accompanied by tangible evidence. If any of your past work is classified or company proprietary, be sure to obtain permission for its release to your state board. Otherwise you may have to omit mention of the work or describe it in general terms only.

Organization of Application Material

Give careful thought to the organization of your application. A neat, well-planned application is itself evidence of a professional. Table 4.1 presents one possible framework that has been successfully used. Each of the three sections shown might be separated in a labeled folder. The content summary pages in folders 2 and 3 outline their contents. Table 4.2 gives a possible outline for the project summary pages for the individual projects. Appendix G shows some sample project summary pages.

Keep in mind that the onus of proof of your eligibility for registration lies with you. The application material that you submit is the main vehicle for this proof. The material must be clear, concise, and, of course, accurate. Your references, as well as past and present supervisors, may be called upon to verify it.

Keep a Copy

Before submitting your completed application material, *copy it for your records*. In case the originals are somehow lost, you will certainly need a backup copy. To help prevent loss, send the application material via certified or registered mail. Some boards keep your application material on file, others will return it after they have made their determination of your eligibility to take the examination.

TABLE 4.1   **Possible Organization of Application Material**

Folder No./Label	Contents
1/Application	Application form
	Application fee
2/Education	Content summary page
	College transcripts
	Copies of college diplomas
	Copies of documents showing other education, licenses, etc.
	Copies of title page and abstract of each academic thesis
3/Experience	Content summary page
	1. Project summary page for project 1 (see Table 4.2)
	Supporting evidence for project 1
	2. Project summary page for project 2
	Supporting evidence for project 2
	●
	●
	●

TABLE 4.2   Contents of Project Summary Pages [1]

Project Number	This should agree with your numbering in the application.
Supervisory Status	Describe the degree of responsibility you had on this project. The role of your supervisor should be made clear, and your part on a team should be spelled out.
Problem and Specific Principles Involved	Describe the project concisely, showing which scientific and engineering principles were involved. Include comments on the results achieved and on your contributions.
Illustrations	Describe the supporting evidence you are submitting. This evidence can include reports, drawings, photographs, publications, and possibly models.

1. See Appendix G for sample pages.

Structure of the Examinations

Everything you have done up to this point has been concerned with proving your eligibility to take the exams. Your state board will now count heavily on how you perform on them.

Exam Material and Type

As you have certainly learned by now in your academic and professional experience, how you perform on an exam depends on how well you know the subject matter *and* on how well you know the structure or format of the exam. The former is obvious. The latter concerns (1) what types of problems you can expect, (2) whether the exams are open-book so that your time spent on memorizing information can be minimized, and (3) the method of grading the exams so that you can pace yourself to maximize your grade.

Since the NCEE exams now are used by most states and other jurisdictions (51 out of the 55), they will be used as the prototype for discussions in this chapter. Even if your state does not use these exams, you should still use them to prepare yourself. In this case, of course, you will also need to obtain and study past samples of your state's exams.

Only a few states require personal appearances or oral exams. These are now normally reserved for engineers seeking registration under grandfather clauses (which permit registration based primarily on long experience) and for engineers from non-accredited schools. Generally, if you are prepared for the written exams, however, you should have no trouble with any personal appearance. You should remember the following points if you do receive an invitation to meet the board:

- The board is looking for stable, competent, honest professionals. Dress conservatively and neatly. A coat and tie for men, a dress or suit for women are definitely in order.
- Do not allow the board to confuse you. They may be using pressure interview techniques. Simply answer their questions directly, using reference to basic principles and letting your experience show.
- Be prepared for questions on professional ethics.
- Be prepared to face a board with no representative of your special field, especially if you are registering in one of the less traditional engineering disciplines.

Open-Book Exams

The exams are usually open-book. Be sure you ascertain whether your state's are, because this will influence how you prepare for them. One state, Arizona, does not specify whether its exams are open or closed book until just before the exams. The term "open-book" sometimes means only bound books and bound notes (in ring binders, for example), not loose notes and papers. Thus, if you prepare notes as suggested in the next chapter, be sure to bind them somehow.

The Exams Are Long

The exams are brutally long. Each is eight hours. When taken together, the EIT and PE exams require sixteen hours and cover two consecutive days, with two four-hour sessions each day. As mentioned above, taking the EIT and PE exams separately is preferable; however, in states where it is permitted, many choose to take both exams together. In any case, it is important to be ready for the exams not only mentally but also *physically*.

Exam Structure

The structures of the exams are shown in Tables 5.1, 5.2, 5.3 and 5.4. Carefully note the way to get the maximum grade on each exam, since this will determine which problems you work and how you budget your time. The total grade for each is the sum of the grades on the morning and afternoon sections. Since the EIT exams are machine-graded, you may receive the results of these more quickly than you do those of the PE exams. Although each state determines what a passing grade is, grades above 70 percent on each day's exams are normally passing.

New Format for PE Exam

Starting with the NCEE exams for November 1978, a very significant change in format is being introduced for the PE (Principles and Practice of Engineering) exam. Heretofore, all candidates were given the same set of exam problems from which each selected the problems he or she wished to work. Under the new format, and subject to any variations a particular state might impose,[1] a candidate will take either of two different exams (the choice may be either the candidate's or the board's, depending on the particular state):

- *The "combined" exam,* consisting of a single exam booklet covering the four traditional major engineering disciplines, civil, chemical, electrical and mechanical (plus economic analysis).
- *The "specialties" exam,* consisting of another exam booklet covering nine other disciplines (plus economic analysis).

A PE candidate who is, for instance, a civil engineer will in practically all cases be taking the "combined" exam. In many cases he will work only

---

1. In Michigan, for instance, all candidates must take the "combined" exam.

problems in civil engineering. In other cases he might also work problems in the other three major disciplines. He may or may not be required to do the economics problems. In some jurisdictions the choice will be the candidate's. In others it will be determined by his board.

On the other hand, a PE candidate whose specialty is other than one of the four major disciplines will in all likelihood take what we have termed the "specialties" exam. It is also quite likely that he will confine the problems worked to his specialty, but again this is not necessarily so. Here also the choice may or may not be his, since each board makes its own rules.

Check with Your Board

As you can see, with this new format for the PE exam, the board for each jurisdiction will more than ever exercise control over what problems you may or may not, or must or must not work. *It is therefore essential that you contact your board at the very start so that you will know in what disciplines you will need to concentrate your studies.*

Much Information Is Covered

The information in Tables 5.3 and 5.4 was taken from the NCEE "Typical Questions" pamphlets and from information on past NCEE and other examinations. It is clear from even this abbreviated list why a three-month review and preparation period is recommended. Further, you will spend some of your study time taking sample exams and concentrating on your particular field of engineering and you may find the recommended two or three nights a week barely adequate.

TABLE 5.1   **NCEE Exam Structure for Fundamentals of Engineering (EIT)**

EXAM	TYPE/GRADING/MATERIAL
Fundamentals of Engineering (EIT)	Structured, multiple-choice, machine-graded; no partial credit; covers general scientific, engineering, and economic principles.
Morning (4 Hours)	150 individual problems; 100 correct give maximum grade of 50%; no penalty for wrong answers.
Afternoon (4 Hours)	Usually open-book; approximately 20 problem sets, each with 10 related questions; 5 correct sets give maximum grade of 50%; only the 5 sets indicated by candidate are graded; candidate must do problems from at least 4 fields and no more than one problem in economics.

TABLE 5.2 **New Structure for NCEE Principles and Practice of Engineering (PE) Exam starting November 1978**

EXAM	TYPE/GRADING/MATERIAL
Principles and Practice of Engineering (PE)	Unstructured; answers written in blank solution pamphlet; partial credit given; covers specific applications of engineering principles in a current professional situation; directed toward a person with engineering degree and four years experience in his field.

"Combined" Exam Option

Morning session (4 Hours)	Nine problem sets in each of the 4 major disciplines (civil, chemical, electrical, mechanical) plus one in economic analysis; you will work 4 problem sets according to proctor's instructions; 4 correct problems give maximum of 50%; only the 4 problems indicated by candidate are graded.
Afternoon session (4 Hours)	Same structure as morning session

IMPORTANT: it is entirely at the discretion of each state board as to which problems candidate can elect to do. Thus, you may be allowed to select any 4 problems regardless of discipline, or you may be required to select your 4 problems exclusively from your specialty or you may or may not be required (or allowed) to do the problem in economic analysis. It is therefore essential that you check with your state board well in advance.

"Specialties" Exam Option (other than the 4 majors)

Morning session (4 Hours)	Nine problems in each of 9 disciplines plus one in economic analysis; 4 correct problems give maximum grade of 50%. Only the 4 problems indicated by candidate are graded.
Afternoon session (4 hours)	Same structure as morning session

IMPORTANT: it is entirely at the discretion of each state board as to which problems candidate can elect to do. Thus you may be required to select your 4 problems exclusively from your specialty or you may be allowed to select any 4 problems regardless of discipline (highly unlikely) or you may or may not be required (or allowed) to do the problem in economic analysis. It is therefore essential that you check with your state board well in advance.

TABLE 5.3   **Main Areas Covered by NCEE
Fundamentals of Engineering (EIT) Exam**

EXAM	TOPICS
Fundamentals of Engineering (EIT)	
Morning session	Chemistry
	Computer Science
	Dynamics
	Economic Analysis
	Electrical Theory
	Fluid Mechanics
	Mathematics
	Materials Science
	Mechanics of Materials
	Nucleonics
	Statics
	Systems Theory
	Thermodynamics
Afternoon session	Computer Science
	Dynamics
	Electrical Theory
	Engineering Economics
	Fluid Mechanics
	Materials Science
	Mechanics of Materials
	Statics
	Systems Theory
	Thermodynamics, classical
	Thermodynamics, statistical

Table 5.4   **Main Areas Covered by NCEE Exam for Principles and Practice of Engineering (PE), starting November 1978**

EXAM	TOPICS	REMARKS
PE "Combined" Exam Option		
Morning session	Chemical Civil Electrical Mechanical	In some states candidate may be allowed to take it as a truly "combined" exam and may select any 4 problems regardless of discipline; in other states candidate must stick to his or her major discipline. It is therefore essential that candidate contact the state board well in advance to learn what will be required.
Afternoon session	Same	
PE "Specialties" Exam Option		
Morning session	Manufacturing Ceramic Industrial Petroleum Agricultural Nuclear Sanitary Structural Aeronautical/Aerospace	Candidate must stick to problems in his or her specialty. (In any particular jurisdiction, some departure from this restriction is theoretically possible but highly unlikely.)
Afternoon session	Same	

Preparing for the Examinations 6

Preparation is where you will be spending most of your effort. This chapter is designed to help you make the time you spend preparing fruitful.

Use Exam Preparation Checklist

It is important to develop and use a schedule for covering each of the areas in which you will be examined (See tables 5.3 and 5.4). The exam subject area preparation checklist in Appendix B is a convenient place to write down your schedule dates. In allotting time for each topic, remember that your purpose is to review and/or learn the general material in each discipline. You will not have time to become an expert in all areas, and indeed that is not required to pass the exams. You must be well versed in your own specialty, however.

The Mark of an Engineer

Working many sample problems is the key to successful preparation. A mathematician or physical scientist might be able to solve all the problems on the exams if given enough time to derive all the pertinent information from basic concepts. The mark of the engineer, however, is his ability to proceed directly to a result without having to rediscover the necessary relationships. He has these at his fingertips from having solved many similar or related problems in the past. It is just this immediate fluency that the exams test because of their strict time limits.

Overprepare

Therefore, your best preparation is to *work problems, work problems, work problems.*

Condensed Information Sheet (CIS) Notebook

To provide a focal point for your reviews and to have something to review and to use during the examinations, you should prepare a notebook of condensed information sheets (CIS). As mentioned earlier, any such notes which you plan to take into the exams should be bound in some fashion. During your review of each subject, you should condense the material so that it will fit on one or two sheets. You will then be able to find important equations, relationships, concepts, and so forth quickly. It is also useful to have a CIS on conversion factors and useful constants. This CIS notebook has the advantageous effect of reducing the time you will spend leafing through reference books during the exams.

Because of the way the exams are structured and scored, you can omit studying some areas (see Tables 5.1-5.4). But there are two areas in which you are expected to be competent and in which you may have to make some extra effort. These are your own engineering discipline and engineering economic analysis.

Economic Analysis

Economic analysis is important to all engineers, and there will be questions covering it. (Check in advance with your board whether or not you will be required to do the problems in this area on the PE exam.) You should be able to handle present and future worth, compound interest, mortgages, sinking funds, equipment depreciation and retirement, break-even analysis, annual cost, rate of return, selection of alternative approaches, and similar calculations. These areas, while mathematically simple, are often unfamiliar to young engineers. Be sure to prepare a CIS on this area containing the pertinent money-time relationships. Also include interest tables in your references for the exams if your electronic calculator does not handle such financial work.

Do Not Slight Your Engineering Specialty

There is a tendency while studying to treat your own area of specialization as simply another topic, especially when you think you are up to date after several years of practice. Most probably, you are fluent in only a narrow area, and there may be no exam questions covering this area! Therefore, you must review the principal concepts in your whole field. If your area is one of the relatively new ones to PE exams, such as nuclear engineering, you will be unable to obtain books especially geared for the exams and containing problems used on past exams. In that case, you must rely on your own ability to select likely problems for review and on the few past exams you can obtain. On the other hand, in older fields such as electrical, mechanical, and civil engineering, you will find many excellent exam preparation texts. (See Appendix D.)

Other Preparation Aids

Local universities, the National Society of Professional Engineers (NSPE), technical schools, and even correspondence schools offer PE exam review courses. These are certainly useful. They all rely on your own persistence and self-preparation, however, and, if you are motivated and organized, you can probably do as well on your own.

How to Study

Figure 6.1 is a flowchart of a recommended study approach. It provides for

- a systematic review of each area
- working many sample problems
- feedback for more review when deficiencies are found
- making up the Condensed Information Sheets
- working sample exams under actual conditions
- deciding on what equipment and which references you will need in the exams.

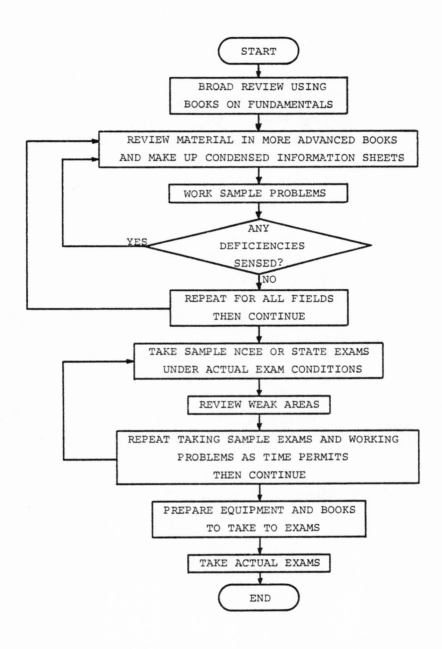

FIGURE 6.1 EXAM PREPARATION FLOWCHART

Note from the flowchart that a general, broad review is recommended at the beginning using a text such as Polentz's *Engineering Fundamentals for Professional Engineers' Examinations* (see Appendix D). This will help you identify your weak areas early. Then you can move on to the more specialized references.

One of the important milestones in the registration checklist (Appendix A) is writing to the NCEE and your state to obtain past exams. Also, check with PEs you know to see if they can provide you with exams they may have obtained during their own preparation.

Keep Sample Exams Unopened Until Ready for Them

When you receive the sample or past exams, do not look through them! Wait until you have prepared for taking them, and take them under realistic conditions, including time limits. Most probably, you will not receive full past exams for the EIT but only those for the PE. The Typical Questions pamphlets published by NCEE, however, give an idea of the EIT exams.

Do not feel in any way guilty about obtaining past exams for study. It is a perfectly legitimate means of exercising and focusing your problem-solving ability. The state boards and NCEE encourage their use, and few candidates forgo the assistance they provide.

Note Most Useful References

During your reviews, note which references you consult most frequently and include them in your book checklist in Appendix C.

Once you have worked through the procedure shown in Figure 6.1, you should be well prepared. The procedure should expose you to a variety of problems, and it is this broad exposure that will best prepare you for the examinations. You will probably even find that this wide-ranging review is enjoyable and that you will learn a great deal of useful engineering.

Taking the Examinations

Ways to Maximize Scores

There are several ways for you to maximize your scores in addition to preparing for the exams by reviewing the subject material thoroughly:

- bringing the right equipment to the exams
- being prepared physically to take the exams
- being familiar with the exam format
- making optimum use of the allotted time during the exams.

Do Not Take Too Many Books

Appendix C helps you decide what equipment and references to take to the examinations. You should carefully plan what you will take. A few states place a limit on the number of books that can be brought into the examination room, but most do not. Do not take so many books that you will exhaust yourself carrying them around or that you will be tempted to spend too much time leafing through them during the exams. Instead, take the ones with which you are most familiar and which cover each topic. Also, rely on your **CIS** notebook.

Get a Motel Room

Give yourself every opportunity to be fresh and ready for the examinations. If the exams are not given in your city (and maybe even if they are), plan to spend the night before each exam in a motel near the examination center. Dress comfortably, and bring amenities to make yourself comfortable. Allow plenty of time to avoid the added stress of being late, hunting for parking space, not finding a suitable seat, and so forth.

Restrictions on Calculators

Electronic calculators have virtually displaced the slide rule as the engineer's problem-solving tool. State boards have accommodated to this and permit use of these calculators in the exams subject to facilities they have available for electrical power. Some permit battery-powered units only, others provide electrical outlets and even recommend that long extension cords be brought along.

There are several restrictions on calculators, however. Most states allow only silent calculators that will not disturb others while they are being used. Another interesting restriction is just starting to be invoked by several states: the use of programmable calculators is not permitted. Thus, be sure to verify that the calculator you plan to use will in fact be allowed.

EIT Exam, Morning

It is not likely that you will complete all of the problems on the exam given on the first part of the EIT but it is recommended that you try to. It is likely, however, that you will get the maximum grade (50%). There are 150 short problems, and a half-point is given for each correct one. No points are subtracted for incorrect answers. With 240 minutes (four hours) you could allot about 2½ minutes per problem. Use your time efficiently and try to mark the best answer to *every* question.

EIT Exam, Afternoon

The exam on the afternoon portion is a little more difficult. Each problem set has multiple questions. Additionally, you must select only five problem sets as those to be graded. Each of the problem sets has ten multiple-choice questions; and all ten parts of all five problem sets must be correct to earn the maximum score (50%). Thus, getting the maximum grade means working correctly all parts of all problems you select.

Allot some of your available time to deciding which of the approximately twenty problem sets you will work. Note that you may be required to work "at least one" or "no more than one" economics problem. With a total of four hours available, you might allot forty minutes for each problem set, leaving forty minutes for deciding which of the sets to work. Of course, the best strategy is to find the easiest problems quickly and work these. Then more than forty minutes will probably be available for the more difficult problems.

**PE Exam,
Morning and
Afternoon**

The morning and afternoon exams on the PE exam are similar. They are not as structured as those of the EIT; there are no multiple-choice options, and the answers are written in blank exam answer pamphlets rather than on machine-gradable sheets. You may not be allowed any loose scratch paper and have to use the left-hand pages of your exam booklets for this. You will have four hours in the morning and four hours in the afternoon.

You will want to work problems from your area of specialization, although there may be no restrictions on which problems you work. Again, your state may require that one of the problems you work be an economics problem.

For each exam *read the instructions carefully and think about what they say.*

In your professional area, there will probably be nine problems, of which you must work four. You will have about fifty minutes per problem, plus about forty minutes for deciding which problems to work. Here too your best strategy is to do the easiest problems first so that you can spend any extra time on the difficult ones. This approach gives quite a psychological boost since you realize that you have completed a significant portion of the exam. You may even finish the exam before you encounter any difficult problems.

Because of the time limitations, write the answers to your problems only once. On the other hand, you should perform enough preliminary scratch work to demonstrate that your rationale is correct. Your work will consequently not be as neat as it would be with plenty of time. To offset this somewhat, you can highlight your principal assumptions, equations, and results with colored pencils or felt pens.

Earn Your Partial Credit

Unlike the EIT exams, *partial credit* is given on the PE exams. Therefore, give the grader every opportunity to understand you. It is a good idea to start each problem by stating the essential problem and describing how you will proceed to a solution. Give assumptions and reference equations you do not derive by citing a text and page number. If you run into numerical problems that you cannot explain within the time limitations, explain what the problem is and where the probable error or misunderstanding lies. Again, communicate your solution clearly to the grader from beginning to end.

Completing Registration 8

Pay Remaining Fees

After notification that you have passed your exams, there are some little things that must be done to complete your registration as a PE. The most important is to pay the state board any remaining registration fee. You will then receive a registration card, a registration certificate, and probably a roster of all persons registered in your state. You are also authorized to purchase a state-approved seal (about $15) and rubber stamp (about $10), which you will use to certify your professional work.

Obtain Seal and Stamp

Review Your New Status

With your registration complete, you are legally empowered to carry out your profession. At this time it is good to reread the state laws and guidelines concerning your authority, responsibility, and professional conduct. They will now seem more meaningful.

Registration in More than One Jurisdiction

Many engineers are involved with national companies that have projects throughout the United States and find they need to be registered in more than one state or jurisdiction. As mentioned earlier, registration is a right reserved by the states and there is no such thing as national registration. Consequently, an engineer must register in each state in which he wishes to practice.

No Need To Repeat Process

Fortunately, this does not mean that it is necessary to repeat the arduous process required for initial registration. Engineering organizations have slowly succeeded in achieving a great deal of uniformity in registration requirements among states, primarily through nearly unanimous adoption of the NCEE uniform examinations. An engineer who was initially registered on the basis of these exams thus finds that he has satisfied one of the main requirements for registration in each state.

Equivalent Requirements

Most state boards now grant to engineers seeking registration in their states credit for "equivalent" requirements that have been met in a previous registration. Thus, if the state board normally requires the NCEE exams and you have passed the exams in another state, all that is required to complete the subsequent registration is to submit evidence of previous registration, experience, and education on the state's standard application, accompanied, of

Registration by Comity

course, by the necessary fees. This is called registration by comity. It is relatively painless except for the engineer who has obtained registration under a "grandfather clause" or during a grace period before written exams were introduced in a state. Since he will not have satisfied comparable or equivalent requirements, he will probably have to take the written examinations to secure the subsequent registration unless he can satisfy the board of his professional "eminence." Thus, there is a definite advantage to having the initial registration based on the written examinations, especially if multiple registration will be sought later.

NEC

Although there is no national registration or licensing of engineers, there is a national "certification" program, National Engineering Certification (NEC), conducted by NCEE. The goals of this program are

- to base certification on requirements so stringent that all states will accept them
- to maintain a record on all certificate holders, which is updated at least once yearly

- to serve as a trustworthy clearinghouse for information on certificate holders for subsequent registrations.

The requirements for obtaining the NEC certificate are

- current registration in some state based on the written examination
- graduation in an engineering or related science curriculum
- no license revocations except for nonpayment of renewal fees
- qualification under one of the following:

Minimum Education	Minimum Age	Written Exam	Professional Experience
ECPD grad	30 yrs	16 hrs	12 yrs
Non-ECPD grad	40 yrs	16 hrs	20 yrs
ECPD grad	40 yrs	8 hrs	20 yrs

Clearly these requirements are such that not every engineer registered in a state is eligible for the NEC certificate. For applications and full information on the NEC program, obtain "Circular of Information 1000" from the NCEE.

Temporary Permits

As an alternative to obtaining full registration in other states after initial registration, engineers can obtain temporary permits in all but twelve jurisdictions.[1] These permits are issued to registrants from other states and limit engineering work to a specific project for a year or less.

State borders still represent barriers to the practice of professional engineering, however. The programs of the NCEE and the efforts by state boards should continue to lower these barriers, although a national engineering registration program does not seem likely in the foreseeable future. Multiple registration will thus remain a necessity for some.

Canadian Registration

A brief review of the requirements for registering in Canada is given here. Registration in Canada is also as a "Professional Engineer" (but abbreviated "P. Eng."). The eleven provinces and territories are the licensing jurisdictions; and the provincial and territorial "associations" and "corporations" of professional engineers perform the licensing function of the U.S. state boards.

Canadian and U.S. engineers seeking full Canadian registration are treated almost identically. The U.S. engineer, however, has to obtain "landed immigrant status" in Alberta, British Columbia, and Quebec. Those seeking registration in Quebec must also obtain Canadian citizenship and show a working knowledge of French. All Canadian jurisdictions except the Yukon Territory require establish-

Residency

---

1. The twelve: Alaska, Arizona, California, Colorado, Connecticut, Guam, Michigan, Missouri, Minnesota, Tennessee, Vermont, Virginia.

ing residence. Because of these citizenship and residency requirements, U.S.-based engineers normally obtain a temporary permit rather than full registration for projects in Canada. Temporary Canadian permits are discussed below.

Similar Educational Requirements

Educational requirements for Canadian registration are similar to those for U.S. registration, and all Canadian jurisdictions accept Canadian as well as ECPD-approved curricula in the areas in which they grant registration.

Experience Requirements Somewhat Less

Professional experience requirements tend to be only two years rather than the four years required in the United States. This minimum experience must usually have been obtained in either Canada or the United States, although Quebec requires that one of the years be spent in Canada.

Examinations Not Normally Required

The Canadian examinations are used more for qualifying nonuniversity-trained engineers than as a general requirement for all candidates to complete. This differs markedly from the U.S. practice. NCEE examinations and certificates are not formally recognized in Canada, but they are given serious consideration.

Thus, the qualified university-trained engineers will normally only be required to fill out an application, submit evidence of experience, and pay the fee to secure registration. Nonuniversity-trained engineers are required to have about five years' experience and pass an examination, or have about eight years' experience.

Temporary Permits in Canada

All Canadian jurisdictions grant temporary permits (up to a year) to practice engineering. The following ones will base such permits solely on registration in the United States: New Brunswick, Newfoundland, Nova Scotia, Prince Edward Island, Quebec, and Saskatchewan. The remaining jurisdictions take into account education and experience in addition to other registration. No written examinations or interviews are required for those already registered elsewhere.

Manitoba and Quebec require holders of temporary permits to collaborate with a P.Eng. registered in the province. All Canadian associations will renew their temporary permits beyond a year. Alberta will issue temporary permits to firms and corporations in addition to individuals.

An engineer already registered in Canada seeking either temporary or full registration in other Canadian jurisdictions normally only need submit an application, fee, and certificate of good standing from his or her home association.

Registration of Foreign Engineers in the U.S.

For those already registered in Canada or another country, the process of registration in a U.S. jurisdiction is somewhat more complex. Since candidates registered in other countries have satisfied such differing requirements, it is difficult for a state to ascertain just what an individual's qualifications are and whether equivalent requirements have been met. Some countries complicate the situation by making it difficult to obtain their registrants' records.

Most states grant temporary permits to foreign engineers on the basis of registration in their home country. This alternative is generally adequate.

When full registration is sought, however, most foreign candidates, even those registered in their home countries, will find themselves in the position of the unregistered U.S. candidate. Indeed, since a foreign candidate will probably have attended a non-ECPD school and since some states require U.S. citizenship or actions toward acquiring it, the foreign candidate may face even more difficulties. On the other hand, some requirements may be lifted if he can show to the state board's satisfaction that he has met equivalent requirements elsewhere. Completion of written exams (eight to sixteen hours) is often basis for being excused from the EIT exam and occasionally from the PE exam.

Thus, it is important for the foreign candidate to present his qualifications as completely as possible to the registration board in the state of interest. The board always is the final authority on what is acceptable and on what must be completed.

Appendices

Appendix A   Registration Checklist   59
Appendix B   Exam Subject Area Preparation Checklist   61
Appendix C   Checklist of Equipment and Books to Take
                  to the Examinations:
                      Checklist of Equipment   63
                      Checklist of Books for EIT Exam   65
                      Checklist of Books for PE Exam   67
Appendix D   References:
                      General   69
                      Professional Ethics   74
                      Economic Analysis   75
                      Structural and Civil Engineering   76
                      Mechanical Engineering   76
                      Electrical Engineering   77
Appendix E   Addresses:
                      Engineering Organizations — General   79
                      NCEE Participating Organizations   80
                      State Boards   82
                      Canadian Associations   90
Appendix F   Stated Positions of Some Engineering
                  Societies Regarding Registration   92
Appendix G   Sample Project Summary Pages   99

Appendix A **Registration Checklist**

Action	Completion Date Relative[1]	Actual[2]	Completed[3]
Write state board for registration package	−20	_____	[ ]
Write schools for transcripts	−18	_____	[ ]
Complete application, have notarized, and submit with fee via registered mail	−15	_____	[ ]
Write NCEE and state for past exams	−14	_____	[ ]
Obtain access to references and start working through exam preparation procedure (Figure 6.1)	−12	_____	[ ]
Receive notification that you can take exams	− 8	_____	[ ]
Reserve motel room near exam center	− 4	_____	[ ]
Finish working through exam preparation procedure	− 1	_____	[ ]
Collect items on checklists of equipment and books to take to the exams (Appendix C)	− 1	_____	[ ]
Travel to exam center	0	_____	[ ]
Take exams	0	_____	[ ]
Receive notification of having passed exams and complete registration	+16	_____	[ ]

---

1. Weeks prior to or following exams.
2. Fill in actual date to be completed.
3. Check when completed.

Appendix B Exam Subject Area Preparation Checklist

REVIEW AREA	WEEK(S) OF	COMPLETED
MATHEMATICS	_____	[ ] check
CHEMISTRY	_____	[ ] when completed
MECHANICS	_____	[ ]
MATERIALS	_____	[ ]
FLUID MECHANICS	_____	[ ]
THERMODYNAMICS	_____	[ ]
ELECTRICITY AND ELECTRONICS	_____	[ ]
NUCLEONICS	_____	[ ]
HEAT TRANSFER	_____	[ ]
ECONOMIC ANALYSIS	_____	[ ]
PROFESSIONAL ETHICS	_____	[ ]
YOUR OWN FIELD: _____	_____	[ ]
OTHER TOPICS:		
_____	_____	[ ]
_____	_____	[ ]
_____	_____	[ ]

Appendix C **Checklists of Equipment and Books to Take to Exams**

Equipment to Take to Exams

Letter(s) admitting you to the examinations . . . . [   ]

#2 pencils (used for mark sensing
on the machine-graded parts); at lease 6 . . . . [   ]

Large eraser . . . . [   ]

Electronic calculator, extra batteries, charger,
extension cord (check well in advance to see
whether electrical outlets will be available) . . . . [   ]

Slide rule (for backup or
if no calculator is used) . . . . [   ]

Scratch paper, assorted graph
paper (may not be allowed) . . . . [   ]

Straightedge . . . . [   ]

Watch . . . . [   ]

Colored pencils and/or felt pens . . . . [   ]

Sweater or jacket (the exam room may be cold) . . . . [   ]

Light snacks for midmornings and midafternoons
(to keep blood sugar level up) . . . . [   ]

Aspirin (just in case) . . . . [   ]

Suitcase and/or briefcase
(to carry books and equipment) . . . . [   ]

Books for EIT Exam   *fill out and refine during preparation*

AREA	BOOKS
GENERAL	
MATHEMATICS	
CHEMISTRY	
MECHANICS	
MATERIALS	
FLUID MECHANICS	
THERMODYNAMICS	
ELECTRICITY AND ELECTRONICS	
NUCLEONICS	
HEAT TRANSFER	
ECONOMIC ANALYSIS	
OTHER AREAS:	

Books for PE Exam   *fill out and refine during preparation*

AREA   BOOKS

GENERAL   _____

                 _____

                 _____

ECONOMIC ANALYSIS   _____
(if you are required, or
permitted, to do problem(s)   _____
in this area)
                          _____

                          _____

YOUR ENGINEERING FIELD _____

                 _____

                 _____

OTHER AREAS:

_____   _____

                          _____

                          _____

_____   _____

                          _____

                          _____

_____   _____

                          _____

                          _____

Appendix D **References**

Most of the reference works listed here were specifically written as aids for preparation for the EIT and PE examinations. They cover the fundamentals, as well as the traditional engineering disciplines. The newer, derivative disciplines are less thoroughly covered and other references will have to be utilized. Prices are shown where known, although they are, of course, subject to change.

General

How to Become a Professional Engineer, by John D. Constance, PE, McGraw-Hill, 1966 ($16.50)

An interesting and comprehensive text on registration, although some significant changes have occurred since this revision. (A new edition is scheduled for late 1978.)

Engineering Fundamentals for Professional Engineers' Examinations, by Lloyd M. Polentz, PE, McGraw-Hill, 1961 ($16.85).

An excellent review book especially for EIT examinations, containing reviews of important concepts and typical questions and answers in mathematics, mechanics, fluid mechanics, thermodynamics, mechanics of materials, electricity and electronics, chemistry, and engineering economics. This book should be worked through from cover to cover.

Professional Engineer's Examination Questions and Answers, 3d ed. by William S. LaLonde, Jr., PE, and William J. Stack-Staikidis, McGraw-Hill, 1976 ($16.50).

This is one of the most widely available general review texts. Two hundred problems make up the very good review of basic fundamentals and can be used in preparing for the EIT examination. Four hundred additional problems cover chemical, civil, electrical, mechanical, and structural engineering, as well as engineering economy and land surveying. The detailed solutions to those 600 problems are given in a section separate from the questions.

The Practice of Engineering in the United States, NCEE, 1973 (free).

This short pamphlet summarizes quite well the broad subject of registration. It defines the practice of engineering, discusses state registration and national certification, and summarizes state registration requirements. Except for a general inflation in the registration fees, the requirements shown are still valid.

A Case for Professional Registration, NSPE Publication No. 2210, 1972 (free).

A brief review of some of the reasons to secure legal registration as a PE.

Selected Bibliography on Professional Engineers and Engineers-in-Training Examination Questions, NSPE Publication No. 2201, 1975 (free).

A listing of useful publications for EIT and PE exam preparation. Also, a list of some of the available correspondence courses is given.

Typical Questions brochures: *Fundamentals of Engineering* and *Principles and Practice of Engineering,* NCEE, 1974 ($1.00 each).

These two excellent pamphlets should be obtained by each person planning to take the NCEE exams. They give sample problems for all sections of the exams. *Principles and Practice of Engineering* is one of the few sources of sample problems in some of the newer engineering disciplines.

Hydraulics Refresher ($4.00), *Thermodynamics Refresher* ($5.00), *Mechanical Refresher* ($4.25), and *Electrical Refresher* ($4.25), all by John D. Constance, PE, available directly from the author at 625 Hudson Terrace, Cliffside, N.J. 07010.

These books review pertinent material and expose the candidate to EIT multiple-choice questions and PE subjective questions.

Professional Engineering Examinations, vol. I (*1965-1971*), NCEE ($16.50).

Solutions/Professional Engineering Examinations, vol. II (*1965-1971*), NCEE, 1974 ($33.50 for solutions only; or $47.50 with above volume of questions).

These rather expensive volumes provide the greatest concentration of past examination questions that the candidate is likely to find. Vol. III, due out in 1978, will eventually replace Vol. I.

1777 Review Problems: From EIT and Engineering Examinations with Answers and Typical Solutions, by R. C. Brinker et al., International Textbook, 1967 ($10.00).

This book is divided into sections on fundamentals, applied engineering, and exam questions. The first two sections contain questions and answers illustrating the range of questions usually asked. The third section is presented as an actual exam with questions only. Answers are given separately in an appendix.

NOTE: The NCEE has scheduled for late 1978 publication a compendium of all state laws concerning registration of engineers, land surveyors and architects.

Circular of Information No. 1000, Regarding NCEE Services, Council Records, and Certification, NCEE (free).

Presents details of the National Engineering Certification program.

Engineer-in-Training License Review, 8th ed., by C. Dean Newnan, Engineering Press, 1976 ($8.95).

This is an outstanding problem practice book consisting of questions and answers in all areas of the EIT exams.

Engineering Fundamentals: Principles, Problems, and Solutions, by D. G. Newnan and B. E. Larock, Wiley, 1970 ($18.25).

An excellent text for EIT review with typical questions and solutions.

Engineers' P.E. Application Kit, by John D. Constance, available from author (625 Hudson Terrace, Cliffside Park, N.J. 07010), 1975 ($10.00)

A do-it-yourself package for writing up your application and avoiding pitfalls. Sample application records.

Fundamentals of Engineering, by James W. Morrison, Arco, 1978, 2 vols ($17.50 each)

Home study for EIT exam featuring study materials, exercises, sample problems, solutions. Vol 1 reviews math, chemistry, physics, economics, electrical; vol 2 reviews statics, dynamics, strength of materials, materials science, thermodynamics, fluid mechanics.

EIT Engineer-in-Training Examination, by James W. Morrison, Arco, 1978 ($10.00)

Features 600 practice questions with solutions.

Professional Ethics

Ethics for Engineers, NSPE Publication No. 1105, 1974 (free).

This brief pamphlet contains the code of ethics developed by the NSPE Ethical Practices Committee. It also references various subjects to the code and to NSPE's three-volume *Opinions of the Board of Ethical Review.*

Legal Aspects of Engineering, by Richard C. Vaughn, Prentice-Hall, 1962.

The chapter on ethics comments on state registration and professionalism and also includes the widely used Canons of Ethics and Rules of Professional Conduct.

Economic Analysis

Engineering Economics for Professional Engineers' Examinations, 2d ed., by Max Kurtz, PE, McGraw-Hill, 1975 ($14.95).

This book is a must for review by the engineer who has had little experience in engineering economics. It can be worked through for exposure to the types of questions most often asked in this area. It addresses pertinent mathematics, the time value of money, sinking funds and annuities, valuation of bonds, depreciation and depletion, annual cost, analysis of variable costs, and legal and business aspects of the construction industry.

Principles of Engineering Economy, 6th ed., by Eugene L. Grant and W. Grant Ireson, and Richard S. Leavenworth, Ronald Press, 1976 ($16.00).

This is a standard text in engineering economics. It presents more detail of subjects than normally is needed for the examinations, but is very useful as a reference while studying.

Engineering Economy, 5th ed., by H. G. Thuesen, W. J. Fabrycky, and G. J. Thuesen, Prentice-Hall, 1977 ($16.95).

Structural
and Civil
Engineering

Structural Engineering for Professional Engineers'
Examinations Including Civil Engineering Review,
2d ed., by Max Kurtz, PE, McGraw-Hill, 1968
($16.45).

A very good review for the Principles and Practice of Engineering exam in the structural and
civil areas. Concepts are explained and accompanied by typical examples with solutions.
Areas covered are mechanics of various structural bodies; stability of structures; concrete,
fluid, and soil mechanics; surveying and route
design; and water supply and sewerage.

Civil Engineering License Review, 7th ed., by
Donald G. Newnan, PhD, PE, Engineering Press,
1974 ($13.95 paper, $17.50 cloth).

Mechanical
Engineering

Mechanical Engineering for Professional Engi-
neers' Examinations Including Questions and
Answers for Engineer-in-Training Review, 2d ed.,
by John D. Constance, PE, McGraw-Hill, 1969
($18.65).

This is a comprehensive review book for the
mechanical engineering exam. Through text
and illustrative questions and answers, it reviews mechanics, machine design, gearing,
mechanism, hydraulics, thermodynamics, fuels

and combustion, various types of engines and turbines, pumps, fans, blowers, compressors, heat transmission, refrigeration, heating and ventilating, and air conditioning.

Mechanical Engineering License Review, 2d ed., by Richard K. Pefley and Donald G. Newnan, Engineering Press, 1974 ($13.95 paper, $17.50 cloth).

Electrical Engineering

Electrical Engineering for Professional Engineers' Examinations, 3d ed., by John D. Constance, McGraw-Hill, 1975 ($17.00).

This book provides a good review of the traditional subjects of electrical engineering. Material is reviewed and illustrated by questions and answers. Direct and alternating current measurements, components, and machinery are discussed, as well as lighting and electronics.

Electrical Engineering License Review, 4th ed., by Lincoln D. Jones and Donald G. Newnan, Engineering Press, 1975 ($11.95 paper, $15.50 cloth).

Appendix E   **Addresses**

Engineering Organizations — General

NCEE   National Council of Engineering Examiners
              P.O. Box 5000
              Seneca, S.C. 29678

              NCEE produces the national exams and makes available
              past exams and a variety of other pertinent information.

NSPE   National Society of Professional Engineers
              2029 K Street
              Washington, D.C. 20006

              NSPE offers a wide range of inexpensive publications of
              interest to engineers. It also has correspondence courses for
              the EIT and PE exams.

ECPD   Engineers' Council for Professional Development
              345 East 47th Street
              New York, N.Y. 10017

              ECPD publishes a list of institutions with accredited curri-
              cula leading to degrees in engineering.

Addresses continue on following pages

NCEE Participating Organizations

ACSM	American Congress on Surveying & Mapping 210 Little Falls Street (P.O. Box 601) Falls Church, Va. 22046
ACEC	American Consulting Engineers Council Suite 410-413, 1155 - 15th Street, N.W. Washington, D.C. 20005
AIChE	American Institute of Chemical Engineers 345 East 47 Street New York, N.Y. 10017
AIIE	American Institute of Industrial Engineers 25 Technology Park/Atlanta Norcross, Ga. 30071
AIME	American Institute of Mining, Metallurgical & Petroleum Engineers 345 East 47 Street New York, N.Y. 10017
ANS	American Nuclear Society 244 East Ogden Avenue Hinsdale, Ill. 60521
ASAE	American Society of Agricultural Engineers 2950 Nils Road St. Joseph, Mich. 49085
ASEE	American Society for Engineering Education One DuPont Circle, Suite 400 Washington, D.C. 20036
ASCE	American Society of Civil Engineers 345 East 47 Street New York, N.Y. 10017

ASME   American Society of Mechanical Engineers
                345 East 47 Street
                New York, N.Y. 10017

CCCELS   California Council of Civil Engineers and
                    Land Surveyors
                811 Forum Building, 1107 - 9th Street
                Sacramento, California 95814

CSPE   California Society of Professional Engineers
                Central Office, P.O. Box 9005
                Sacramento, California 95816

IEEE   Institute of Electrical & Electronics Engineers
                2029 K Street, NW
                Suite 605
                Washington, D.C. 20006

MSRLS   Michigan Society of Registered Land Surveyors
                P.O. Box 344
                Lansing, Mich. 48902

NICE   National Institute of Ceramic Engineers
                65 Ceramic Drive
                Columbus, Ohio 43214

NSPE   National Society of Professional Engineers
                2029 K Street, N.W.
                Washington, D.C. 20006

SME   Society of Manufacturing Engineers
                20501 Ford Road
                Dearborn, Mich. 48127

State Boards   Alabama State Board of Registration for Professional
Engineers and Land Surveyors
64 North Union Street, Room 606
Montgomery, Alabama 36130

Alaska State Board of Registration for Architects,
Engineers & Land Surveyors
Pouch D
Juneau, Alaska 99811

Arizona State Board of Technical Registration
1645 W. Jefferson Street
Suite 315
Phoenix, Arizona 85007

Arkansas State Board of Registration for Professional
Engineers & Land Surveyors
Box 2541
Little Rock, Arkansas 72203

California State Board of Registration for
Professional Engineers
1006 4th Street, 6th Floor
Sacramento, California 95814

Canal Zone Board of Registration for Architects and
Professional Engineers
Box 223
Balboa Heights, Canal Zone

Colorado State Board of Registration for Professional
Engineers and Land Surveyors
1525 Sherman, Room 600B
Denver, Colorado 80207

Connecticut State Board of Registration for Professional
Engineers and Land Surveyors
State Office Building, Room 533
Hartford, Connecticut 06115

Delaware Association of Professional Engineers
1508 Pennsylvania Avenue
Wilmington, Delaware 19806

District of Columbia Board of Registration for
Professional Engineers
614 H Street, N.W., Room 109
Washington, D.C. 20001

Florida State Board of Professional Engineers and
Land Surveyors
6990 Lake Ellenor Drive, Suite 100
Orlando, Florida 32809

Georgia State Board of Registration for Professional
Engineers and Land Surveyors
166 Pryor Street, S.W.
Atlanta, Georgia 30303

Guam Territorial Board of Registration for Professional
Engineers, Architects and Land Surveyors
Department of Public Works
Government of Guam
P.O. Box 2950
Agana, Guam 96910

Hawaii State Board of Registration of Professional
Engineers, Architects, Land Surveyors and Landscape
Architects
P.O. Box 3469
Honolulu, Hawaii 96801

Idaho State Board of Engineering Examiners
842 La Cassia Drive
Boise, Idaho 83705

Illinois Department of Registration & Education
628 E. Adams Street
Springfield, Illinois 62786

Indiana State Board of Registration for Professional
Engineers and Land Surveyors
1021 State Office Building
100 N. Senate Avenue
Indianapolis, Indiana 46204

Iowa State Board of Engineering Examiners
State Capitol Complex
821 Des Moines Street
Des Moines, Iowa 50319

Kansas State Board of Engineering Examiners
535 Kansas Avenue, Room 1202
Topeka, Kansas 66603

Kentucky State Board of Registration for Professional
Engineers and Land Surveyors
University Station, Box 5075
Lexington, Kentucky 40506

Louisiana State Board of Registration for Professional
Engineers and Land Surveyors
1055 Saint Charles Avenue, Suite 415
New Orleans, Louisiana 70130

Maine State Board of Registration for Professional
Engineers
State House
Augusta, Maine 04330

Maryland State Board of Registration for Professional
Engineers & Professional Land Surveyors
1 South Calvert Street, Room 802
Baltimore, Maryland 21202

Massachusetts State Board of Registration of
Professional Engineers and Land Surveyors
Leverett Saltonstall Building, Room 1512
100 Cambridge Street
Boston, Massachusetts 02202

Michigan State Board of Registration for Professional
Engineers & Land Surveyors
1116 S. Washington Avenue
Lansing, Michigan 48926

Minnesota Board of Architecture, Engineering, Land
Surveying and Landscape Architecture
Metro Square Building, 5th Floor
St. Paul, Minnesota 55101

Mississippi State Board of Registration for Professional
Engineers and Land Surveyors
P.O. Box 3
Jackson, Mississippi 39205

Missouri Board for Architects, Professional Engineers
and Land Surveyors
P.O. Box 184
Jefferson City, Missouri 65101

Montana Board of Professional Engineers and
Land Surveyors
Lalonde Building
Helena, Montana 59601

Nebraska State Board of Examiners for Professional
Engineers and Architects
512 Terminal Building
941 "O" Street
Lincoln, Nebraska 68508

Nevada State Board of Registered Professional Engineers
P.O. Box 5208
Reno, Nevada 89513

New Hampshire State Board of Registration for
Professional Engineers
4 Park Street
Concord, New Hampshire 03301

New Jersey State Board of Professional Engineers
and Land Surveyors
1100 Raymond Boulevard
Newark, New Jersey 07102

New Mexico State Board of Registration for Professional
Engineers and Land Surveyors
P.O. Box 4847
Santa Fe, New Mexico 87502

New York State Board for Engineering and
Land Surveying
99 Washington Avenue
Albany, New York 12230

North Carolina State Board of Registration for
Professional Engineers and Land Surveyors
1307 Glenwood Avenue, Suite 152
Raleigh, North Carolina 27605

North Dakota State Board of Registration for
Professional Engineers
P.O. Box 1264
Minot, North Dakota 58701

Ohio State Board of Registration for Professional
Engineers and Land Surveyors
180 East Broad Street
Columbus, Ohio 43215

Oklahoma State Board of Registration for Professional
Engineers and Land Surveyors
401 United Founders Tower
5900 Mosteller Drive
Oklahoma City, Oklahoma 73112

Oregon State Board of Engineering Examiners
Labor & Industries Building, 4th Floor
Salem, Oregon 97310

Pennsylvania State Registration Board for
Professional Engineers
Box 2649
Harrisburg, Pennsylvania 17120

Puerto Rico Board of Examiners of Engineers,
Architects and Surveyors
Box 3271
San Juan, Puerto Rico 00904

Rhode Island State Board of Registration for
Professional Engineers and Land Surveyors
20 State Office Building
Providence, Rhode Island 02903

South Carolina State Board of Engineering Examiners
710 Palmetto State Life Building
Columbia, South Carolina 29201

South Dakota State Board of Engineering and
Architectural Examiners
2040 West Main Street, Room 212
Rapid City, South Dakota 57701

Tennessee State Board of Architectural and
Engineering Examiners
550 Capitol Hill Building
301 7th Avenue, N.
Nashville, Tennessee 37219

Texas State Board of Registration for Professional
Engineers
1400 Congress, Room 200
Austin, Texas 78701

Utah Representative Committee for Professional
Engineers and Land Surveyors
330 E. Fourth South Street
Salt Lake City, Utah 84111

Vermont State Board of Registration for Professional
Engineers
Cabot Science 143
Norwich University
Northfield, Vermont 05663

Virginia State Board for the Examination and
Certification of Architects, Professional Engineers
and Land Surveyors
P.O. Box 1-X
Richmond, Virginia 23202

Virgin Islands of the United States Board for
Architects, Engineers and Land Surveyors
Public Works Department
P.O. Box 476
St. Thomas, Virgin Islands 00801

Washington State Board of Registration for
Professional Engineers and Land Surveyors
Division of Professional Licensing
P.O. Box 649
Olympia, Washington 98504

West Virginia State Registration Board for
Professional Engineers
1800 E. Washington Street, Room 411
Charleston, West Virginia 25305

Wisconsin State Examining Board of Architects,
Professional Engineers, Designers and Land Surveyors
1400 E. Washington Avenue
Madison, Wisconsin 53702

Wyoming State Board of Examining Engineers
Barrett Building
Cheyenne, Wyoming 82002

Addresses continue on following pages

Canadian   Canadian Council of Professional Engineers
Associations   Suite 401
                      116 Albert Street
                      Ottawa, Ontario K1P 5G3

Constituent Provincial Associations:

Association of Professional Engineers of Alberta
215 One Thornton Court
Edmonton, Alberta T5J 2E7

Association of Professional Engineers of British Columbia
2210 West 12th Avenue
Vancouver, British Columbia V6K 2N6

Association of Professional Engineers of Manitoba
177 Lombard Avenue
Room 710
Winnipeg, Manitoba R3B 0W5

Association of Professional Engineers of New Brunswick
123 York Street
Fredericton, New Brunswick E3B 3N6

Association of Professional Engineers of Newfoundland
P.O. Box 8414
Postal Station A
St. John's, Newfoundland A1B 1X1

Association of Professional Engineers of Nova Scotia
P.O. Box 129
1828 Upper Water Street
Halifax, Nova Scotia B3J 2M4

Association of Professional Engineers of Ontario
1027 Yonge Street
Toronto, Ontario M4W 3E5

Association of Professional Engineers of Prince Edward Island
92½ Queen Street
P.O. Box 278
Charlottetown, P.E.I. C1A 4B1

Order of Engineers of Quebec/Ordre des Ingenieurs du Quebec
1100-2075 Rue University
Montreal, Quebec H3A 1K8

Association of Professional Engineers of Saskatchewan
Suite 220
2220 Twelfth Avenue
Regina, Saskatchewan S4P OM8

Association of Professional Engineers of Yukon Territory
c/o 5 Kalzas Place
Whitehorse, Yukon Territory

Appendix F Stated Positions of Some
Engineering Societies Regarding Registration

AIChE

American Institute of Chemical Engineers:

"All state legislatures now require registration (licensing) of all engineers who offer their services directly to the public. Also, many regulatory agencies, courts of law, clients and employers require or recommend registration as evidence of engineering competence and the number that do so is increasing. Registration is one avenue to define an engineer's level of competence and it does provide assurance to the public that those with the title "Registered Professional Engineer" or "Licensed Professional Engineer" have met certain minimum requirements at one point in their careers.

In consideration of these facts, the AIChE recommends that its members become registered in their respective states as soon as possible after entering the profession.

In making this recommendation, the AIChE recognizes that registration is no substitute for individual professional responsibility for engineering work, a responsibility which is based upon adequate knowledge, experience and moral integrity and which extends throughout the entire career of every engineer.[1] "

ASCE

American Society of Civil Engineers:

(Policy approved by Board of Direction, October 27, 1973) "The American Society of Civil Engineers has actively, and continually, supported and encouraged the registration of professional engineers since 1909, when the first committee was formed to prepare a workable registration program plan.

The primary function of the registration process is to protect the public, and the public interest, from persons who may be uneducated or incompetent in the practice of

---

1. Policy statements to reflect AIChE's position on relicensing, industrial exemptions (which at this time AIChE does favor), and the form of business organization are under consideration and will be issued in the future.

engineering — this protection includes the health, safety, convenience, and welfare of the public in matters pertaining to engineered works.

The American Society of Civil Engineers endorses, supports, and promotes the registration of Professional Engineers, as being in the best interests of the public.

In support of this policy, the American Society of Civil Engineers:

1) Endorses in principal the National Council of Engineering Examiners Model Law (1972 Revision) as an acceptable guide for the promulgation of registration laws within the jurisdictions which have legal authority to regulate.

2) Recommends that the appropriate regulating body in each jurisdiction adopt the "Model Rules of Professional Conduct" as approved by the National Council of Engineering Examiners in August of 1972.

3) Supports the activities of the National Council of Engineering Examiners (NCEE) in the coordination and guidance of engineering registration procedures.

4) Encourages all Sections and District Councils to cooperate with other engineering organizations and statewide Engineering Councils in the advancement of professional engineering registration.

5) Advocates that persons in public service who hold responsible positions in engineering design, management, construction, operations, supervision,

and other professional level engineering activities should be Registered Professional Engineers.

6) Urges that those engineering faculty members who teach engineering subjects should be Registered Professional Engineers.

7) Recognizes that a large number of corporations presently offer and render engineering services to the public and recommends that the protection of the public be assured by requiring that such corporations satisfy the following requirement:

   a) That the individual engineers in responsible charge of the practice of engineering shall be registered in the State having jurisdiction over performance of such practice.

8) Recommends that all States should license by Comity those registered engineers in good standing who have the following qualifications:

   a) Graduation from an ECPD accredited curriculum.
   b) 4 or more years of satisfactory experience in engineering practice.
   c) Satisfactory completion of 16 hours of approved examinations.

9) Recommends that Registered Professional Engineers actively assist and support their Registration Boards, and that they actively seek out and support the candidacy of qualified Professional Engineers to fill positions on these Boards.

10) Recognizes that the registration function is to serve the interest of the public and recommends:

a) That Registration be maintained at the State level.
b) That control of registration policies and procedures, and the administration of same, be vested in qualified Registered Professional Engineers.
c) That registration and renewal fees be consistent with the cost of the services involved in the administration and enforcement of the appropriate regulatory act.

This policy is intended to insure that the regulation of the engineering profession is directed to the protection of the public health, safety, and welfare. The registration of those professional engineers who possess the expected knowledge, judgment, and integrity will provide the public with appropriate service.

The right and duty of Government to protect its citizens through the police power of the state is recognized. The registration, or licensing, of engineers is an extension of the police power, and it should be exercised in those areas where a need for the protection of the citizenry can be shown."

The above are stated positions in full of two major societies. It would be impractical here to give the full statements of all societies who have issued them. Instead we present below summaries or excerpts from their statements:

ACSM   "Endorses in principle the NCEE Model Law (1972 Revision) as an acceptable guide for the promulgation of registration laws within the jurisdictions which have legal authority to regulate."

AIIE   "It is imperative that professional registration be recognized as a major activity. . . . The Board of Trustees has, on numerous occasions, approved the adoption of the Model Law."

AIME   Recognizes registration as professional engineers to be a desirable action to be taken by its members and encourages its members to register.

ASME   Approves in principle the NCEE Model Law.

IEEE   In 1976, the IEEE Board adopted a 10-point statement on registration that included recommendations on the legal restrictions, on standards for examinations, on the identification of licensed practitioners, on industrial exemptions, and on educational support. Membership responses are being incorporated into a revision that presumably will be presented to the IEEE Board for adoption before the end of 1978.

NSPE   "NSPE believes that the state engineering registration laws should apply to all engineers responsible for engineering design of products, machines, building, structures, processes, or systems that will be used by the public; and urges management to engage only registered professional engineers for responsible engineering positions. Management is also urged to promote registration among all of its engineering staff.

The Society is opposed to proposals to exempt engineers in industry from the state registration laws and recommends the phasing out of existing exemptions in state registration laws."

SME   "SME is strongly supportive of Professional Registration and feels that the industrial exemption should eventually be removed.

It is the position of SME that the industrial exemption should not be removed quickly and arbitrarily."

Appendix G   Sample Project Summary Pages

On the following pages we see that portion of a typical application form which shows the employment record, followed by samples of a few of the related project summary pages . . .

EVIDENCE OF EDUCATION AND EXPERIENCE
(STATEMENTS SHOULD BE PLAINLY PRINTED OR TYPEWRITTEN)

1. Preliminary education St. Johns Country Day School, Orange Park, Fla, 1959
   (Name of high or preliminary school) (Years attended) (Year of graduation)

2. Study of engineering satisfactorily completed: (see Attachment A)
   University of Florida, Nuclear Engineering, 9/63-10/64, M.S. in Nuclear Engineering
   MIT   Nuclear Engineering, 9/65-12/68, Ph.D. in Nuclear Engineering
   (Name of University, College or Technical School) (Curriculum pursued) (Date of attendance) (Date of Graduation) (Degrees)

NOTE: A college or engineering school graduate shall submit an official certificate of his graduation; if not a graduate, applicant shall submit transcripts of any collegiate or engineering study completed.

3. Record of experience or lawful practice in chronological order.

Engagement Number	DATE Month and Year From	DATE Month and Year To	STATE IN ORDER: EMPLOYMENT RECORD — (a) Name, Location and Character of Business of Employer (b) Kind of Work Done By Applicant (c) Degree of Responsibility Therefor	TIME (Years and Months) Pre-engineering work	TIME (Years and Months) Engineering work	TIME (Years and Months) Total Time	Name and address of someone familiar with each employment record, preferably the person to whom applicant reported.
1	summer of 1958		Jacksonville Blowpipe Company, Jacksonville, Fla. Manufacturer of Industrial blowers and grinders. Built devices for measuring characteristics of Industrial centrifugal fans, pony brake, dynamic and static pressure devices, flow restrictors, and ran tests to generate fan performance curves.	3m		3m	W.T.S. Montgomery, President Jacksonville Blowpipe Co., Jacksonville, Fla.
2	summers of 1960 and 1961		Overhaul & Repair Department, USNAS, Jacksonville, Fla. Overhaul and repair facility for Navy aircraft. Engineering Aide, mainly involved with plant equipment layout work and similar industrial engineering tasks.	6m		6m	?
3	summers of 1962 and 1963		Pan American World Airways, Guided Missiles Range Div., Cape Canaveral Fla. Range Contractor for Atlantic Missile Range. Summer Trainee. Worked in missile range performance evaluation and non-destructive testing. The latter involved X-ray radiography, radiograph developing and interpreting.	6m		6m	?
4	summer of 1967		Argonne National Laboratory, Argonne, Illinois. National Laboratory. Resident Associate (staff of the University of Chicago). Developed and implemented methods for calculating fast reactor		3m	3m	Dr. A. Travelli, Argonne National Lab. Argonne, Illinois

		analysis (mainly sodium void calculations).			
5	9/67 8/67 - .	MIT (Massachusetts Institute of Technology, Cambridge, Mass.) University. Teaching Assistant, helped professors teach courses in Advanced Reactor Physics and Reactor Engineering Lab, gave lectures, developed and graded homework.	9m		9m	Drs. I. Kaplan and M. Driscoll, Dept. of Nuclear Eng'ing, MIT, Cambridge, Mass. 02139
6	3/69 12/71	Combustion Engineering, Windsor, Connecticut. Manufacturer of Nuclear and Fossil Powerplants. Sr. Staff Physicist, directed R&D project which developed a 3-D reactor design system now in routine use as an engineering design tool;developed xenon stability control algorithms; performed work in fuel cycle analysis and powerplant optimization.	2y9m	2y9m		Dr. R. Lee Combustion Engineering 1000 Prospect Hill Rd. Windsor, Ct 06095
7	1/72 present	Singer, Simulation Products Div., Silver Spring, MD. Manufacturer of simulation equipment ("Link" trainers). Sr. Staff Engineer and later Mgr. of Reactor Systems. Developed reactor design capability at Singer; organized, staffed, trained, personnel, and directed new section responsible for simulation of reactor cores, control rod systems, nuclear instrumentation systems, safeguard systems, emergency core cooling, containment, and other systems. Involved in hardware and software tasks. Supervised 25 people from clerical to Ph.D. engineers. Involved in bid and proposal preparation, technical sales presentations, and negotiations.	2y5m	2y5m		S. T. Kremzner Singer 2121 Industrial Pkwy Silver Spring, MD 20904
		TOTAL TIME (ADD EACH COLUMN SEPARATELY)	2y	5y5m	7y5m	

ENGAGEMENT NUMBER: 6, Part 2

SUPERVISORY STATUS: The plotting work reported here was all done
by myself, with no assistance from my supervisors or from those
working for me.

PROBLEM AND SPECIFIC PRINCIPLES: Difficult problems are often made
easier if they can be visualized. The need arose in my mind at
Combustion Engineering for visual projections of three dimensional
data of reactor core performance. I wrote a computer program to
do this plotting on a digital drafting machine. The problem is
to take three dimensional data and project it on a plane according
to the aspect chosen by two Euler angles. A complication is that
lines behind objects between them and the point of observation must
be deleted, the hidden line problem. This problem was solved by
having the computer investigate every point in the three dimensional
space and compare it with a "fence" that is updated with each point
and which blocks the field of view. Two data points, one of which
lies below the fence, are not connected: only the line segment
from the point in view to the fence is drawn. The program proved
very successful and found innumerable applications including the
following:
    a. visualization of core flux and power distributions for designing
       flat power distributions
    b. visualization of fuel assembly power distributions for designing
       a properly zone-loaded assembly, especially in the case of
       Plutonium recycling
    c. visualization of xenon oscillations for finding adequate control
       procedures
A memorandum written by me is attached which illustrates one of the
uses under c.

ILLUSTRATIONS: The attached memorandum combines two of my interests
at Combustion Engineering, computer plotting and control of xenon
oscillations in Pressurized Water Reactors. The code described
above (TRIDPLOT) and developed by me is used to generate the
figures numbered 2 through 9. The memorandum is self-explanatory.
A memorandum describing the program TRIDPLOT has also been attached.

ENGAGEMENT NUMBER: . 6, Part 4

SUPERVISORY STATUS: In this piece of work, I assisted Mr. Paul
Zmola in an economic cost analysis of the reactor industry. My
main contributions were in the area of statistical analysis of
the data. Mr. Zmola was in charge of the project and prepared a
paper based on our work.

PROBLEM AND SPECIFIC PRINCIPLES: Powerplant manufacturers as well
as electric utilities like to keep track of trends in the industry-
wide costs of electrical generation. Total plant costs vs. net
electrical capacity and unit electrical costs ($/KW) vs. net
electrical capacity are two important numbers. Variation of
costs with time is also a significant indicator. Several of these
indices were determined for 41 light water reactor plants in the
United States. Various functions (linear, semi-log, etc.) were
used to fit statistically the information on the 41 plants.
Further economic work was done jointly with Mr. Zmola including
nuclear fuel cycle analysis and a total plant optimization study.

ILLUSTRATIONS: Attached is a copy of Mr. Zmola's paper's Abstract
and Summary along with the Acknowledgement of my contribution. The
paper was presented in Vienna in 1971.

ENGAGEMENT NUMBER: 7, Part 1

SUPERVISORY STATUS: During this piece of work I was Sr. Staff
Engineer at Singer's Simulation Products Division. I was supervised
by a Section Head who had no reactor experience, and I essentially
served as Singer's resident nuclear expert. The work was done
independent of any technical supervision.

PROBLEM AND SPECIFIC PRINCIPLES: In the simulation of nuclear
powerplant reactor cores, it is necessary to account in great
detail for a large amount of information covering normal operation
and accident conditions. Singer basically required a reactor design
calculational capability in order to generate this data. I developed
this capability in part by collecting reactor design codes from
national laboratories and in part by writing new codes (computer
programs). One of these is the REACTOR code which is used for efficient
calculation of three dimensional power shapes with the effects of
control rods, fuel temperature and moderator temperature feedbacks,
soluble boron, the iodine and xenon chain, thermohydraulics of coolant
flow through the core, fuel burnup, and others. The code is based
on published work that I performed earlier at Combustion Engineering
and on a new technique for solving the higher order difference
equations. The code has been in use at Singer almost two years
and during that time several simulator customers became familiar with
it and wanted to use it for their own calculations of core performance.

ILLUSTRATIONS: A copy of the REACTOR User's Manual is attached.

ENGAGEMENT NUMBER: 7, Part 2

SUPERVISORY STATUS: same as Part 1 plus being Principal Investigator
on R&D task which produced the work

PROBLEM AND SPECIFIC PRINCIPLES: The problem was the development of
a technique for simulating the behavior of a reactor core (PWR),
accounting for all the complexity visible to a reactor powerplant
operator. Due to the large number of in-core instrument systems,
the simulation model had to deliver a large amount of data, and
since the model had to run in real-time with all other powerplant
system models, the constraints on computer resources were severe.
Thus, where a full design calculation for one point in time might
run several hours on a CDC-6600, the simulator model had to run
in about 150 msec every second on a considerably smaller computer.
I applied a technique which has enjoyed success in numerical
approximation work and which was related to the work I had performed
on my doctoral dissertation. This was the modal synthesis approach
where a large amount of known information can be factored into the
approximation of the detailed neutronics equations, and less
reliance is placed on the mathematics to calculate all details.
On the other hand, a large amount of detail results from the model.
The exact details of the model are considered proprietary by Singer,
and I am not at liberty to discuss them here. I simply state that
the model is being used on reactor powerplant simulators under
construction by Singer.

ILLUSTRATIONS: none

Index

Age requirements, for registration, 8, 11
Alabama, xiii
Alaska, 52n
Alberta, 52
American Academy of Environmental Engineers, xvi
American Association of Cost Engineers, xvi
American Congress on Surveying and Mapping, on registration, 96
American Institute of Chemical Engineers, on registration, 93
American Institute of Industrial Engineers, on registration, 97
American Institute of Mining, Metallurgical and Petroleum Engineers, on registration, 97
American Institute of Plant Engineers (AIPE), xvi
American Production Inventory Control Society (APICS), xvi
American Society of Civil Engineers, on registration, 93-96
American Society for Nondestructive Testing (ASNT), xvii
American Society for Quality Control (ASQC), xvi
American Society of Safety Engineers (ASSE), xvi
American Welding Society (AWS), xvii
Application
  copying, for personal records, 20-21
  form, described, 17
  organizing material for, 20-21
  package, components of, 17
  sample project summary pages, 98-104 (Appendix G)
Arizona, 25, 52n
Associations, professional
  addresses of, 79-81 (Appendix E)
  See also Technical societies

Boards of registration, state, xi, xv, xviii-xix, 8, 21, 23, 47
  addresses of, 82-89 (Appendix E)
  discretion over exam problems worked, 27, 29 (Table 5.2)
  personal appearance before, 24
  powers of, 56
Books
  checklists of, 63-65 (Appendix C)
  reference, 69-77 (Appendix D)
  use of, during exam, 41
  See also Notebooks; Notes; Reference books
British Columbia, 52

Calculator, electronic, 35
  programmable, 42
  restrictions on use of, 42
California, 52n
Canada
  citizenship requirement for registration, 52-53
  education/experience requirements for registration, 53-54
  "landed immigrant" status, and registration, 52
  licensing jurisdictions, 52
  professional associations, addresses of, 90-91 (Appendix E)
Certification, xv-xvii
Character references, 7
  selection of, 18
Checklists, use in exam preparation, 33
"Circular of Information 1000" (NEC), 50
Citizenship
  and exams, 12
  requirements for Canadian registration, 53
  U.S., required in some states, 55
Colorado, 52n
Comfort, physical, and exams, 41-42

Comity, registration by, 49
Condensed information sheets (CIS), 34-36, 41
Connecticut, 52n
Constants, 34
"Consumer Protection Agency," xiv
Conversion factors, 34

Deadlines, for exam application, 13
Department of Justice, U.S., xiv
Dress, for board interview, 24

Economic analysis, 26, 35
Economics, problems on PE exam, 44
ECPD, *see* Engineers' Council for Professional Development
Education
  ECPD-approved curricula, 8
  professional experience as substitute for, 8-9, 53
  requirements for registration: in U.S., 7; in Canada, 53-54
  verification of, 17, 18, 19
EIT, *see* Fundamentals of Engineering exam
"Eminence," professional, as alternative to exams, 11, 50
Engineer in Training, 4, 9-10
  See also Fundamentals of Engineering exam
Engineering Fundamentals for Professional Engineers' Examinations (Polentz), 38
Engineering societies, *see* Associations, professional; Technical societies
Engineering Technology Certification Institute, xvi
Engineers' Council for Professional Development (ECPD), 8
Environmental Protection Agency (EPA), xii, xiii
Equal employment opportunity commission, xiv
Equipment, for exams
  checklist of, 63 (Appendix C)
  felt pens, 45
  pencils, 45
  See also Calculator, electronic
Ethics, professional, 24
Examinations
  "combined," 26, 29 (Table 5.2)
  EIT, described, 42-43

length of, 25
machine-graded, 44
multiple-choice, 43
NCEE, 7, 13, 28 (Table 5.1), 29 (Table 5.2), 49, 50, 53
open-book, 23, 25, 28
oral, 8, 24
PE, described, 26-27
review courses for, 36
scheduling of, 10-11
"specialties," 26-27, 29 (Table 5.2)
structure of, 25, 28 (Table 5.1), 29 (Table 5.2)
written, 8
See also Fundamentals of Engineering exam; Principles and Practice of Engineering exam
Examiners, state boards of, *see* Boards of registration, state
Exemptions, from registration, xiii-xiv, 1, 2
"grandfather clause," 24, 50
"manufacturer's," xiii-xiv
Experience, preprofessional, defined, 9
Experience, professional
  as alternative for education, 8
  Canadian requirements, 53-54
  defined, 9
  U.S. requirements, 2, 17

Federal Trade Commission, xiv
Fees
  application, 17
  for Canadian registration, 54
  EIT, 10
  for exams, 8, 11
  for state registration, 47
Florida, xv, xviii
Foreign Engineers, registration of, 54
French (language), required for registration in Quebec, 52
Fundamentals of Engineering exam (EIT), 4
  areas covered, 30 (Table 5.3)
  content, 10
  format and structure, 10, 28 (Table 5.1), 42-43
  requirements for, 9-10
  strategy in taking, 42
  when to take, 15

Georgia Institute of Technology, 16n

Grace period, for registration, 50
Grades
  maximum, 28 (Table 5.1), 29 (Table 5.2)
  passing, 26
"Grandfather clause," 24, 50
Guam, 52*n*

Institute of Certifying Engineering Technicians (ICET), xvi
Institute of Electrical and Electronics Engineers, on registration, 97
Interest tables, 35
Intern Engineer, 10
Interviews, personal, 24
Iowa, xviii-xix

Justice, U.S. Department of, xiv

"Landed immigrant" status, and Canadian registration, 52
Licensing, *see* Registration
Licensing, boards of, *see* Boards of Registration, state

Manitoba, 54
"Manufacturer's exemptions," xiii-xiv
Mathematics, and exams, 15-16
Michigan, 26*n*, 52*n*
Minnesota, xviii, 52*n*
Missouri, 52*n*
"Model Law," xiv
Money-time relationships, 35
Montana, xiv
Multiple registration, 7, 49-51, 52-54

National Association of Corrosion Engineers (NACE), xvi
National Certification Commission in Chemistry and Chemical Engineering (NCCCCE), xvi
National Council of Engineering Examiners (NCEE), xii, xiii, xiv, 7, 11*n*, 38
  and multiple registration, 52
  uniform examinations, 7, 13, 49, 50; and Canadian registration, 53; structure of, 28 (Table 5.1), 29 (Table 5.2)
National Engineering Certification (NEC), 50-51
  requirements for, 51

National Society of Professional Engineers (NSPE), 8, 36
  on registration, 97
New Brunswick, 54
New Jersey, xviii
New York, xiii
Newfoundland, 54
Notarization, of application, 18
Notebooks, bound, 25, 34
Notes, 25, 34
  binding of, 25
  use of, during exam, 41
Nova Scotia, 54
Nuclear Regulatory Commission (NRC), xii

Occupational Safety and Health Administration (OSHA), xii
Open-book exams, 23, 25, 28
Organizations, professional, *see* Associations, professional; Technical societies

PE, *see* Principles and Practice of Engineering exam; Professional Engineer
"P. Eng." (Canadian registration), 52, 54
Permits, temporary, 51-52
  in Canada, 53, 54
  jurisdictions not issuing, 52*n*
  state, for foreign engineers, 55
Polentz, Lloyd M., *Engineering Fundamentals for Professional Engineers' Examinations,* 38
Preparation, for exams
  exam review courses, 36
  exam subject area checklist, 61 (Appendix B)
  physical, 41-42
  sample exams, 36
  study flowchart for, 36-37 (Fig. 6.1)
  texts and references for, 36, 38, 39, 69-77 (Appendix D)
  time required, 15, 28
  use of past exams in, 38
  working sample problems, 34
  See also Books; Condensed information sheets; Notebooks; Notes; Reference books
Prince Edward Island, 54

Principles and Practice of Engineering exam (PE exam), xvii, 10
  areas covered, 31
  compared with EIT, 44
  described, 44-45
  new format (Nov. 1978), 26-27
  partial credit in, 44
  strategy in taking, 44, 45
  structure, 29 (Table 5.2)
  review courses, 36
Professional Engineer (PE), described, 1-2
Project summary pages, 22 (Table 4.2), 99-104 (Appendix G)

Quebec, 52, 53, 54

Recertification, xvii-xix
Reference books, 34, 69-77 (Appendix D)
References, personal, 7
  selection of, 18
Registration (licensing)
  advantages of, 2, 3 (Table 1.1)
  Canadian, requirements for, 52, 53-54
  checklist, 59 (Appendix A)
  by comity, 49
  completing, 47
  defined, 1
  documents, 47
  eligibility for, 4, 7-12, 20
  "equivalent" requirements, 49-50; for foreign engineers in U.S., 55
  exemptions from, xiii-xiv, 1, 2, 24, 50
  fees, 11, 47
  of foreign engineers in U.S., 55
  grace period for, 50
  "grandfather clause," 50
  mandatory, xix-xx
  multiple, 4, 7, 49, 50
  need for, 1-2
  positions of professional associations on, 93-99 (Appendix F)
  requirements for, 7-12, 49-56
  state barriers to, 52
  state boards of, *see* Boards of Registration, state
  U.S. citizenship, required for, in some states, 55
  of U.S. engineers in Canada, 52-54

Residency
  and Canadian registration, 52-53
  requirements, for exams, 12
Reviewing, before exams, 15

Saskatchewan, 54
Scheduling
  of exams, 15
  of preparation for exams, 13, 14 (Fig. 3.1), 15-16, 59 (Appendix A)
Scores, exam
  maximizing, 41
  maximum, on EIT exam, 43
Scratch paper, at exams, 44
Seal, professional, 47
Slide rule, 42
Society of Manufacturing Engineers, xvi
  on registration, 97
Society of Packaging and Handling Engineers, xvi
Stamp, rubber, for certifying professional work, 47
Standards Engineer's Society (SES), xvi
State boards, *see* Boards of registration, state
"Sunset Laws," xv

Technical societies
  and certification programs, xv-xvii
  positions on registration, 93-97 (Appendix F)
  See also Associations, professional
Tennessee, 52*n*
Theses, 19
Training, in nonengineering fields, 19

Vermont, 52*n*
Virginia, 52*n*

Wisconsin, xviii
Work, professional,
  certifying, after registration, 47
  classified or proprietary, 20
  examples of, included in application, 17, 19

Yukon Territory, 52